Work 100

神奇的苍蝇

The Incredible Fly

Gunter Pauli

[比]冈特·鲍利 著

[哥伦]凯瑟琳娜·巴赫 绘

姚晨辉 译

上海远东出版社

特别感谢以下热心人士对童书工作的支持：

匡志强　宋小华　解　东　厉　云　李　婧　庞英元
李　阳　刘　丹　冯家宝　熊彩虹　罗淑怡　旷　婉
杨　荣　刘学振　何圣霖　廖清州　谭燕宁　王　征
李　杰　韦小宏　欧　亮　陈强林　陈　果　寿颖慧
罗　佳　傅　俊　白永喆　戴　虹

目录

Contents

一只孤孤单单的老鼠正坐在他的小笼子里，他整天都在一个实验室中工作，人们在这里试图证明他们的新产品是安全的。

一只苍蝇坐到了老鼠旁边的箱子上，问道：“为什么他们还在用你做实验？”

A lonely mouse is sitting in his little cage. He works in a laboratory all day long, where people are trying to prove that their new products are safe.

A fly sits down on a box next to the mouse and asks, “Why are they still using you for tests?”

为什么他们还在用你做实验？

Why are they still using you for tests?

我必须要搞清楚这些产品是否安全

I have to figure out if these products are safe

“哦，你要知道，女人们会抹口红，孩子们要吃药，男人们也会用肥皂洗手，而在这之前我必须要搞清楚这些产品是否安全。”

“太可笑了！你没有必要做这些事情，你懂的。”

“Oh, you know, before women can apply lipstick, or children can take medicine, or men wash their hands with soap, I have to figure out if these products are safe.”

“How ridiculous! There is no need for you to do that, you know.”

“我是懂，”老鼠感叹道，“但许多国家的法律规定，所有这些化学品必须要经过测试，尽管每个人都知道，这是在浪费时间和金钱……也是在浪费我的生命！”

“你知道吗？我说不定是世界上被研究最多的动物之一呢。”苍蝇说道。

“I know,” sighs the mouse, “but in many countries the law states that all these chemicals have to be tested, even though everyone knows that it’s a waste of time and money … and a waste of my life!”

“Do you know that I am probably one of the most studied animals in the world?” asks the fly.

世界上被研究最多的动物之一

One of the most studied animals in the world

参照我飞行的方式制造机器人

Make robots inspired by the way I fly

老鼠哈哈大笑。“你开玩笑吧！”
他说，“人类能向你学习什么呀？人类
不能飞行，他们没有巨大的眼睛，他们使用肌肉的
方式也和你完全不同。”

“你知道的，我可以用我的翅膀尝味道。而且我的整个
身体都覆盖着传感器。现在人类正想参照我飞行
的方式制造机器人呢。”

The mouse bursts
out laughing. “You’ve got to be
kidding!” he says, “What could humans
ever want to learn from you? People don’t
fly, they don’t have huge, big eyes, and the
way they use their muscles is totally different
from the way you use yours.”

“I can taste with my wings, you know. And
my entire body is covered in sensors.
People now want to make robots
inspired by the way I fly.”

“哈哈，他们可能是想弄清楚你这么小的大脑为何能这么厉害。”老鼠打趣说。

“这可不能混为一谈。不过你说得没错，我的大脑的确非常小。不同的是，我可以将它全部利用起来。”

“我曾听到人类声称大黄蜂不能飞，尽管只要他们不嫌麻烦去看一下，他们就会亲眼看到大黄蜂是可以飞的！”

“Ha-ha, they probably want to figure out how you can do so much with such a small brain,” kids the mouse.

“That’s not a kind thing to say. But it is true all the same; I do have a very small brain. The difference is, I use all of it.”

“I have heard people claim that bumblebees can’t fly even though, if they took the trouble to look, they will see with their own eyes that they can!”

不同的是，我可以将它全部利用起来

The difference is, I use all of it

人类可能有硕大的大脑，但没有充分利用

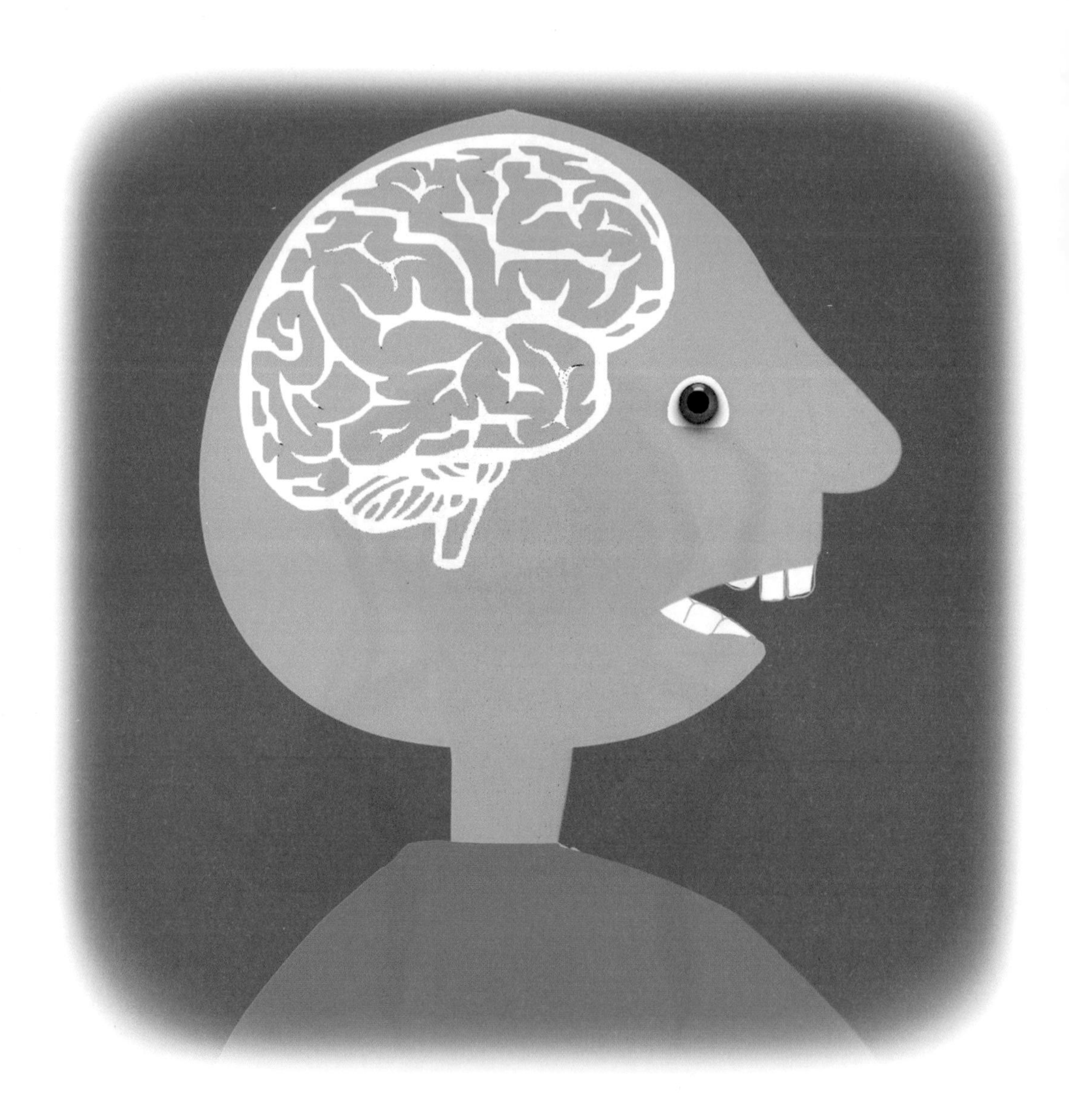

People may have big brains, but don't use them

“人类可能有硕大的大脑，但我认为他们没有充分利用。”

“嗯，他们肯定还没有学会飞行，或作出什么能飞的东西，就像你做的那样。”

“要想飞，你必须要了解风从哪儿来，你还需要有不同的肌肉，为你提供力量或者改变方向。你的大脑必须能够迅速地协调所有这一切，否则你就会一头摔下来，或者更糟——直接飞进捕食者的嘴巴！”

“People may have big brains, but I do not think they use them.”

“Well, they have certainly not yet learned to fly, or make anything fly, as well as you do.”

“To be able to fly, you have to know where the wind is coming from. You also need certain muscles to power you, and other muscles to change direction. Your brain must be able to quickly coordinate all of this or you will fall flat on your face, or worse – fly straight into the mouth of a predator!”

“听起来所有这些对我来说都太难了。”老鼠叹息道。

“我们有些种类的苍蝇可以连续飞10千米，中间不需要进食补充能量。我们甚至可以在空中读取罗盘，辨别方向。”

“All that sounds difficult to me,” sighs the mouse.

“Some of us can fly for up to 10 kilometres without any food for fuel. We can even read a compass and tell direction while we’re up in the air.”

可以读取罗盘，辨别方向

Can read a compass and tell direction

除了南极洲以外，我们可以在任何地方生存

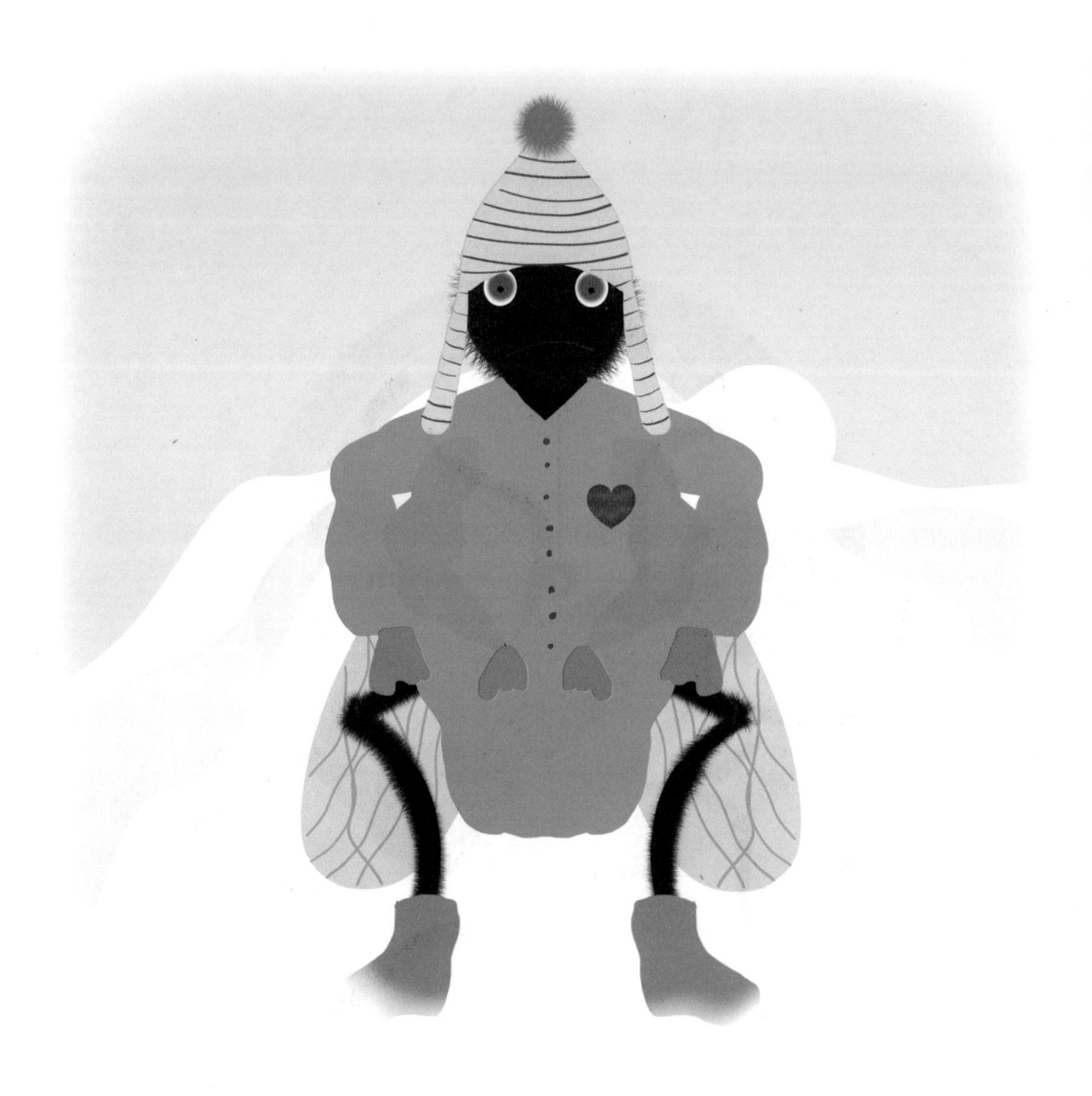

We have learned to live everywhere except Antarctica

“太神奇了！我还以为只有帝王蝶能够做到这一点。”

“永远不要低估一只苍蝇。我们的种类很多，虽然我们身材很小，但除了南极洲以外，我们可以在任何地方生存。”

“你的生活真是丰富多彩，而我和兄弟姐妹们却不得不被关在这儿的笼子里，测试那些化妆品。看这儿，试试这支红色的口红吧。”

“Amazing! I thought only monarch butterflies are able to do that.”

“Never underestimate a fly. There are so many of us and even though we are tiny, we have learned to live everywhere in the world except Antarctica.”

“You lead such an interesting life, while my brothers and sisters and I have to sit here in a cage, testing cosmetics. Here, try on some of this red lipstick.”

“口红？不用了，谢谢！我喜欢自己一身黑的样子。”

“好吧，我喜欢白色。我很高兴自己是一个白化病患者，我希望我的孩子将来也是白色的，但不要红嘴唇！”

……这仅仅是开始！……

“Lipstick? No, thanks! I like being black all over.”

“Yes, and I like being white. I am happy to be an albino, and I do hope that my children will one day be white too, without the red lips!”

... AND IT HAS ONLY JUST BEGUN!...

……这仅仅是开始！……

… AND IT HAS ONLY JUST BEGUN! …

Did You Know?

你知道吗？

More than a 100 million mice and rats are killed in American laboratories because of testing. While the use of dogs and cats in experiments is strictly controlled, the use of mice in laboratories is largely unregulated.

因为做实验，超过1亿只小鼠和大鼠在美国的实验室中丧生。虽然对实验使用的狗和猫有着严格控制，但对于实验室中使用老鼠很大程度上是缺乏监管的。

House mice were the prime reason for taming domestic cats. Nowadays, mice are also kept as pets. The first record of a pet mouse is in the Erya, the oldest surviving Chinese encyclopaedia, which dates back to the 3rd century BC.

对付家鼠是人类驯养家猫的首要原因。如今，小鼠也被作为宠物饲养。有关宠物鼠最早的记录出现在《尔雅》中，那是现存最古老的中文百科全书，其历史可以追溯到公元前3世纪。

Albino animals have a lack of colour pigment in their skin, hair, and eyes. This condition also affects people. There are also albino plants that have lost chlorophyll pigmentation and have white flowers.

白化动物的皮肤、毛发和眼睛中缺乏色素。这种情况也发生在人类中间。同样也有白化植物，它们因为缺乏叶绿素沉着而开白色的花朵。

A mouse has a 1,000 times more neurons than a fly. The fly, however, seems to be able to process information faster than a mouse.

老鼠的神经元要比苍蝇多1000倍，然而，苍蝇处理信息的速度似乎比老鼠更快。

苍蝇每秒钟扑打220次翅膀。在飞行过程中，苍蝇的大脑（包含100 000个神经元）可以在一瞬间对信息进行处理，并迅速改变飞行方向。

Flies beat their wings 220 times per second. While in flight, the fly's brain, consisting of a 100,000 neurons, can process information in a split second and enable it to change course swiftly.

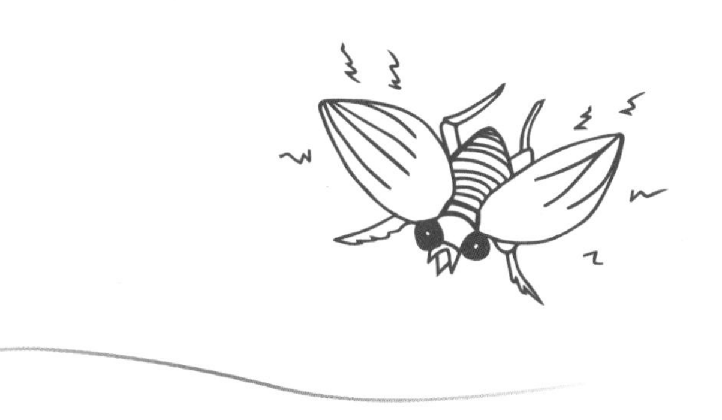

苍蝇并不扇动翅膀，而是创造一个旋涡（类似空气运动的小型龙卷风）来帮助产生升力。苍蝇通过体内的生物陀螺仪来引导它们飞行。

Flies do not flap their wings, but rather create a vortex (small tornado-like air movement) to help generate lift. An internal, biological gyroscope guides their flight.

苍蝇的翅膀上分布着传感器，因此，它们可以用自己的翅膀品尝味道。

Flies' wings have sensors on them and therefore they can taste with their wings.

苍蝇的眼睛是这个星球上速度最快的视觉系统。

Flies' eyes are the fastest visual system on the planet.

Would you like to live in a cage like a mouse, even if you know you are helping people? What other ways of testing the toxicity of products can you suggest people use?

即使你知道自己是在帮助人类，你喜欢像老鼠一样住在笼子里吗？你还可以建议人们使用什么其他方法来测试产品毒性呢？

你觉得苍蝇聪明吗？

Do you think a fly is smart?

Does the size of a brain matter? Or does it matter more how the brain is used?

大脑的体积大小重要吗？还是大脑的使用方法更为重要？

你能想象如果你是一名白化病患者，自己的生活会有什么不同？

Can you imagine how your life would be different if you were an albino?

Let's create a beta movement, an optical illusion where a series of static images on a screen creates the illusion of a smoothly flowing scene. Static images do not physically change but give the appearance of motion because they are rapidly being turned on and off, faster than the eye can follow. This optical illusion is caused by the human eye's inability to process information that is displayed faster than 10 frames per second. Flies are able to process this much better than we can. Now have a look at some LED lights. Even though the lights are individually controlled, our eyes and brain perceive them as a continuous movement of light.

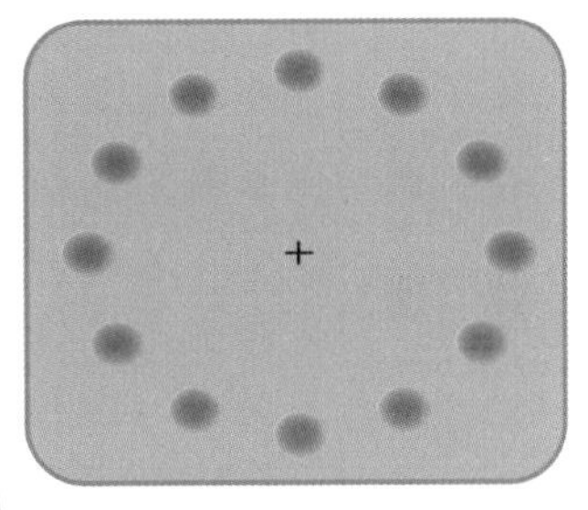

让我们来创造一个β似动现象，这是一种光学错觉，即屏幕上的一系列静态图像产生一个平滑运动场景的错觉。静态图像没有发生物理变化，但给人一种运动的感觉，因为它们被迅速地接通和断开，其速度比眼睛注视的速度更快。这种光学错觉是由于人眼无法处理显示速度大于每秒10帧的信息而引起的。苍蝇在处理这种信息方面比我们人类强得多。现在观察一些LED灯。即使这些灯被单独控制，在我们的眼睛和大脑的感知中，它们也是光的连续运动。

学科知识

Academic Knowledge

生物学	脑细胞在执行不同功能时具有不同的行为；神经元是传输电子和化学信号的细胞；毒性是物质可导致生物体损伤的程度；生物毒素包括细菌和病毒；致癌物和诱变剂；被研究最多的苍蝇是果蝇；帝王蝶及其方向感；老鼠被用作一种方便的实验室研究模型；苍蝇被认为是一个具有复杂和丰富多彩生活的多才多艺的物种，擅长使用其大脑。
化　学	化学毒物包括铅、汞、氢氰酸和甲醇等。
物　理	煤尘、石棉纤维和二氧化硅等物理毒物会干扰肺的生物学过程；β似动会产生光学错觉的效果；似动现象：光脉冲引起的似动。
工程学	复用：如何利用现有的东西做更多事情；通过传感器和生物学指标测量毒性；基于电生理学的工程学：了解苍蝇的视觉航向控制，设计自主飞行机器人（无人机）；四轴飞行器拥有四个旋翼，而不是像直升机那样只有一个；赖卡特探测器（发明于20世纪50年代）是检测运动的简单电路，模仿的是在苍蝇眼睛中发现的神经回路。
经济学	产品需要测试其毒性以获得销售许可证；在广告中使用β似动效果来吸引人们的注意力，诱使他们购买产品；拍摄影视动画时使用β似动技术。
伦理学	在消费品和药品测试中使用动物；利用视觉效果，使大脑认为它看到了一些其实并不存在的东西；展示观察者并没有清醒地意识到但确实存在于大脑中的信息，这种行为在大多数国家是被禁止的。
历　史	出生于瑞士的德国人帕拉塞尔苏斯（原名菲利普·冯·霍恩海姆）在16世纪创立了毒理学学科。
地　理	使用指南针寻找方向，告诉我们朝哪个方向走。
数　学	我们自身的运动知觉是基于视觉、前庭和本体感受的输入，推断运动速度和方向的过程，而移动物体的运动知觉是在给定的某些视觉输入情况下，推断一个视觉场景中物体的速度和方向的过程。
生活方式	我们已经习惯了使用那些对我们的生活和健康有未知影响的化学品，并准备为此承担风险；我们每天都会受到至少3 000条潜意识信息的轰炸，吸引我们的注意力，给我们造成压力，并劝说我们消费产品。
社会学	有毒物质用量：少量的中毒剂量可能是无害的，甚至对健康有益，而大量食用某种无害的物质（如水）也可能会中毒；我们倾向于认为人类具有优越性，只有一个极小大脑的苍蝇是微不足道的，即使它能更有效地使用它的大脑。
心理学	知觉作为诠释周围环境的能力，导致了格式塔心理学的发展；因为一个人的外貌而自豪；追求完美有其优点和局限性。
系统论	我们倾向于关注一个具有两个参数（原因和结果）的函数，并将现实简化为我们自认为理解的样子，然而现实比这复杂得多，认知更多的是模式识别的结果，而不是细节的集合。

情感智慧

Emotional Intelligence

老　鼠

老鼠似乎接受了他的困境。起初，他表露了一点自负的感觉，因为他正在承担一项重要任务——所谓的为人类服务。但他很快就否定了自己扮演的角色，认为他的工作就是在浪费时间和金钱。更糟糕的是，老鼠感觉他是在浪费自己的生命。不过，他仍保持着幽默感。他取笑苍蝇，特别强调她微小的大脑。苍蝇的回答让老鼠质疑了研究苍蝇的实用性。通过对话，老鼠渐渐认识到了自己的局限性，同时表示出对苍蝇更多的尊重。当苍蝇即将离开时，老鼠用自嘲的语气说他会送给苍蝇一些口红。老鼠坦然接受了他作为白化病患者所具有的独特之处，并希望他的孩子能有一个更美好的未来。

苍　蝇

苍蝇对老鼠的困境非常关注。她挺身而出维护自己的权利，并认为老鼠没有受到很好的对待。苍蝇自信满满，并有着很强的自我意识。老鼠的取笑并没有减少她的自豪感，因为她的独特能力也获得了人类的关注，包括神经系统科学家和无人机设计师等。苍蝇花时间解释了她的特性，甚至当这让老鼠感到苦恼的时候，她仍继续在谈论自己，并在结束谈话时再一次确认了她对自己身份的自豪。

艺术

The Arts

让我们观赏一部电影。《蝇王》（THE LORD OF THE FLIES）是一部根据英国作家威廉·戈尔丁（WILLIAM GOLDING）的同名小说改编的电影，戈尔丁因为这篇小说获得了1983年的诺贝尔文学奖。小说讲述了被困在一个无人居住的小岛上的一群男生，试图管理自己，却带来了灾难性后果的故事。和朋友们或一个成年人一起观看电影，并讨论你的感受和见解。

思维拓展

Systems: Making the Connections

大脑有许多独特的功能，不同物种的脑容量各不相同。人们对大脑在过去400万年的发展（包括其大小）进行了详细的研究；老鼠大脑内的神经元比苍蝇多1 000多倍。然而这并不意味着老鼠一定胜过了苍蝇。相反，苍蝇似乎能够更好地利用它的大脑，它对其有限的神经元的有效部署，启发人们改进无人机的设计方式和视觉感知。苍蝇似乎能够非常快地处理信息。它通过眼睛、翅膀上的传感器、脸上的触须捕获信息。它甚至还有一个内置的陀螺仪，来检测方向和确定方位。处理完所有这些信息后，苍蝇以闪电般的速度作出决策和采取行动。而人类的做法是以一种相当线性的方式部署大量的神经元，在处理所有的数据时进行详细的因果分析。苍蝇依托广泛关联的网络或神经元模式，似乎能够更系统地对此进行处理。这意味着，即使神经元的数量有限（如苍蝇仅有100 000个神经元），数学上的模式数量也是无限的。当一只苍蝇逃脱捕食者的抓捕时，其行为不是一个简单的反射，而是一种基于规划、模式识别、快速的数据处理以及迅速纠正措施而采取的行动。科学家们受到苍蝇能力的启发，认识到它为复杂系统的设计提供了大量的机会。苍蝇独特的飞行行为包含了物理和生物科学、流体动力学、复合材料结构以及复杂的非线性数学。对这方面的研究需要一个独特、复杂、跨学科的方法，并需要整合生物学、工程学、物理学和数学。苍蝇可能并不被视作像老鼠那样方便的实验室动物，但至少也是有趣、复杂的昆虫，具有惊人的神经功能。

动手能力

Capacity to Implement

尝试用相机拍摄一张苍蝇的照片，这可能会是一个很大的挑战！拍一张苍蝇头部的特写。现在去当地的宠物商店，拍摄老鼠的照片，同样也拍一张漂亮的头部特写（或者在互联网上寻找苍蝇和老鼠头部的照片）。将这些照片和你父亲或母亲的照片放在一起。现在想象一下，你必须要指出这三者谁的脑子最快。现在，你准备好对三者的大脑进行比较并讨论哪一个更优秀了吗？

故事灵感来自

This Fable Is Inspired by

迈克尔·迪金森

Michael Dickinson

迈克尔·迪金森于1989年在华盛顿大学动物学系获得博士学位，上学期间，迈克尔通过在业余时间做厨师挣学费完成了大学学业，当时他每天早上要做200—300顿饭。在此期间，他开始意识到一个事实，即食物的外观和气味同味道一样重要，从而产生了对感觉的兴趣。在大学里，他的研究最初集中在苍蝇翅膀上感觉细胞的生理学方面。这激发了他对苍蝇的空气动力学和控制飞行所需的电路的兴趣。《科学家》杂志（*The Scientist: Magazine of Life Sciences*）称迈克尔为“苍蝇小子”，因为他将动物学、神经科学和流体力学联系在一起。迈克尔目前在华盛顿大学领导着迪金森实验室，并喜欢用尤克里里琴弹奏爵士乐。

图书在版编目（CIP）数据

冈特生态童书.第三辑修订版：全36册：汉英对照 /
（比）冈特·鲍利著；（哥伦）凯瑟琳娜·巴赫绘；
何家振等译.—上海：上海远东出版社，2022
书名原文：Gunter's Fables
ISBN 978-7-5476-1850-9

Ⅰ.①冈… Ⅱ.①冈… ②凯… ③何… Ⅲ.①生态环
境-环境保护-儿童读物—汉、英 Ⅳ.①X171.1-49

中国版本图书馆CIP数据核字（2022）第163904号

著作权合同登记号图字09-2022-0637号

策　　划 张　蓉
责任编辑 程云琦
封面设计 魏　来李　廉

冈特生态童书
神奇的苍蝇
［比］冈特·鲍利　著
［哥伦］凯瑟琳娜·巴赫　绘
姚晨辉　译